IEC 61850 And Digital Substation Basics

SADANAND PUJARI

Published by SADANAND PUJARI, 2024.

Table of Contents

Copyright 1

About 2

Introduction 3

IEC 61850 Fundamentals 5

The Data Model 13

Generic Substation Events 26

SNTP 31

MMS 37

Ethernet 44

IED 53

Security 60

Use of IEC61850 in Power System Protection 67

Conclusion 72

Copyright

Copyright © 2024 by **SADANAND PUJARI**

All rights reserved. No part of this book may be reproduced, scanned, or distributed in any printed or electronic form without permission. Please do not participate in or encourage piracy of copyrighted materials in violation of the author's rights. Purchase only authorised editions.

IEC 61850 And Digital Substation Basics

This Book Will Provide You With The Basic Knowledge On IEC 61850 And Networking Technologies For Digital Substations

First Edition: Jun 2024

Book Design by **SADANAND PUJARI**

About

IEC61850 Communication Protocol Fundamentals is a comprehensive chapter Book where you can learn industry specific knowledge about the IEC61850 protocol standard such as that found in the utility industry.

I have handcrafted this Book to allow students to acquire core fundamental knowledge from cable component basics to calculating for simple cable installation designs.

I will teach you the basic concepts of the IEC61850 standard as well as the overall IEC61850 Data Model. I will also go over the different major components pertaining to the IEC61850 standard as well as the security aspect of IEC61850 as well as application in line protection within the power utility industry.

If you are a professional who is interested in working in the utility industry or SCADA industry, you will find this Book of great help to get the fundamental knowledge you need to enhance your professional career.

This Book can help you to get an advantage at work or getting a job as it gives you core knowledge in IEC61850 that you would not have otherwise obtained unless you have worked in the industry. And knowledge is power.

So let's get started! Let's start your fulfilling journey and mark an important point of your phenomenal career in this industry!

Introduction

Hello and welcome, In this Book, we will explore the IEC sixty one 850 data model. We will look at the generic substation events such as the GSC and the Ghost Protocol. We will also look at the manufacturing message specification known as AMS, as well as the security features and issues of IEC. Sixty one eight fifty. This Book is for professionals who want to get familiar with the IEC sixty one 850 standard, as well as professionals who work in the implementation of IEC sixty one eight fifty in their data systems. This Book is also for students who want to learn about the communication protocol in general and how they could apply it in their future career.

So now let us look at the Book outline for this particular IEC sixty one 850 Book, we will start off with the chapter fundamentals where we're going to cover the most basic of basic info around the IEC sixty one 850 standard, as well as touching base on the IEC sixty one 850 data model. We will then explore the different components that make up the IEC sixty one, 850 standard. We will start with looking at the generic substation events, then we will move on to the SE and tip. We will then move on to the mess. And then lastly, in this chapter, we will look at the Ethernet layer of the 61 eight 50 standard. Lastly, we will end this Book with our third chapter, which is how is the six to one 850 standard being applied in the real world? First of all, we will look at how 61 one 850 is being implemented in.

And then we're going to look at the security features and issues of IEC sixty one 850. And we will then lastly, look at how this standard is being used in power system protection. Well, I hope that you will learn a lot in this Book if you wish to connect with me. I have a Twitter account with the tech at Boot Camp there. You will learn the latest news in the world of electrical engineering. Also, don't forget to subscribe to my YouTube channel, which is also the tech bootcamp to get the latest free training chapters in which I will share the bits and pieces that I learned throughout my career as an electrical engineer. Once again, thank you for taking interest in this Book, and I'll see you in the next chapter.

IEC 61850 Fundamentals

Let us start off with answering the question, what is IEC? Sixty one eight fifty? Well, IEC sixty one eight fifty is essentially a communication protocol standard for intelligent electronic equipment in electrical substations, of which it is a reference architecture for electric power systems that has been developed by the International Electrotechnical Commission, also known as the IEC. Now the data models described in the IEC sixty one eight 50 are abstract data models that may be translated into a variety of protocols. Existing MMS, which is known as the manufacturing message specification, are included in the standard as well as what we called goose, or the general object oriented substation events and SPV or sampled measured values mappings.

Now, using TCP IP networks or high speed switched ethernet in substation, local area network, or Lance, these protocols can be used to provide the required reaction times as short as below four milliseconds for protective relays within power systems. Now that I've mentioned that IEC sixty one eight fifty is a communication protocol standard. Well, what is exactly a protocol? Well, network protocols provide the standards that allow computers to interact when a physical link has been established. The rules and encoding standards for transmitting data are established by what we call the protocol. This specifies how computers or other devices on a network identify one another. The format in which data should be sent, as well as how the data should be processed once it arrives at its final destination.

Protocols also specify the type of error checking to be used, the data compression method to be used if one is required. How the sending device would indicate that it has completed sending a message, how the receiving device will indicate that it has received a message, as well as how lost or damaged transmissions or packets will be handled. Despite the fact that each network protocol is unique, they all really use the same physical wiring. Multiple protocols can cohabit peacefully over the network medium, and a network architect can utilize similar hardware for a variety of protocols. Because of this common means of accessing the actual physical network, the term protocol independence refers to the ability of devices that are compatible at the physical and data connection levels to run several protocols over the same media.

Now, IEC sixty one 850 is a protocol standard that is commonly used in the energy utility and the manufacturing industry. The three goals of IEC sixty one eight fifty is to number one, unified the information model, number two is the unified server functionality and number three is to unify the file format for the first one IEC sixty one 850 and evils and information model that is compatible with any complying device, including the establishment of a naming hierarchy and certain data structure. They must utilize a consistent structure to create their information and identify the same concept with the same name. This, in turn, decreases the amount of time spent seeking for information, as well as the errors it will generate for the second goal.

IEC Sixty one 850 provides the definition of a communication protocol and a unified server functionality, especially this protocol is the language to all devices in the system speak in order to exchange data, and as a result, it has been developed to meet all requirements for the automation of substations, while also enables interoperability as it is to be implemented by the manufacturers of protection or control devices, skater systems, as well as R2 use or remote terminal units. Lastly, IEC sixty one 850 utilizes an XML based file format with particular criteria to increase automation and configuration capabilities in the engineering process. The main distinct features of IEC sixty one 850 are fast response configured data storage, data modeling. Defined reporting scheme.

Sample data transfer groupings and command types. One of the most distinct features of IEC sixty one in 50 is the fast transfer of event data, that is, this is done via peer to peer communication mode using what we call the generic substation events. This protocol also allows complete storage of configured data under what they call the ACL format. As for data modeling, different standard logical nodes are used to describe primary process objects, as well as protection control functions in the substation. Data and functions linked to a logical device or physical device are stored in logical nodes in this case. IEC sixty one 850 also enables various reporting schemes that can be executed under predefined trigger conditions by a server to client relationships.

Schemes are also defined to support transfer of sampled values from various devices along the system. Lastly, this protocol also allows setting up of group control blocks to handle active groups and various command types that handles and operate various commands securely. The five layers of IEC, sixty one it 50 is what is shown in front of you right now, which are the data model, the layer with which we have the MMS, SMTP, Jesse, the S.V, the mapping layer, the time critical services layer and the Ethernet layer. Now, if you do not know any of the acronyms that I mentioned, don't worry as you will learn all about it throughout this Book.

So what makes this protocol unique? As a result of the IEC, sixty one 850 standard device data integration into information and automation technology has been dramatically improved, resulting in lower cost for engineering and commissioning, as well as for operation and monitoring, diagnostic asset management and maintenance, as well as increased agility. Most earlier utility protocols utilized object models of device functions and device components, while IEC sixty one 850 uses object models for device functionalities, such as using logical devices composed of logical notes and data objects.

Standard data format identifiers and controls are defined in these models, such as those used in the substation and feeder devices, like as measuring units and switches. Standardize behavior for the most common device. Function is specified within these models, which also allow for substantial vendor specialization as a result of these approaches. With multi-vendor interoperability being a possibility. Machines, controllers, graphical user interfaces, computers and data networks work together in industrial control systems or supervisory control and data acquisition systems to automate large scale operations.

It is difficult to combine machines from various manufacturers, which has made interoperability a problem for a lot of these systems. Assume that there are hundreds of production factories dispersed throughout the country, each with their own brand of equipment that communicate in a broad variety. Now what if these plants need to interact with each other in order to work efficiently and effectively? But the protocols that we have were not actually compatible and the devices were not designed to communicate natively. Now in this analogy that I have just said about factories basically is applying to the power grid in general. In this case, substations must interact with one another in order to ensure electricity quality and reliability or dependability.

With so many protocols, however, a great deal of engineering effort needs to be done to make it happen. So what makes IEC sixty one eight fifty an enabler of interoperability? Well, in the past, protocols such as DNP three and Marta's were essentially what we called signal oriented. What this means is that the protocol references data points such as point one zero five oh one zero five zero three from device five. What does this mean? Well, to figure out what this point is, you actually need to check the manual of each device, which is actually quite cumbersome when you have a big, gigantic system like the power grid IEC.

Sixty one eight 50, as we will see, have a standardized nomenclature that is very easy to understand with a defined nomenclature structure and common abbreviations in place. It would be natural to grasp what the point is as long as one is familiar with these standards. As a result, engineering costs will be lower as less work will be spent on finding the relevant points. Consequently, an engineer in charge of the control software does not have to manually input the points and instead the control software can actually automatically detect and identify all IED data points. Lastly, in this chapter, I want to show you the different chapters or the different documents that you will see within the IEC sixty one eight 50 standard. Now this Book, due to the meatiness of the standard, we will not actually cover all of the documents that we see within the standard itself.

And as such, I would highly recommend you after finishing this Book to actually get a copy of IEC sixty one eight fifty, just to get familiar with the standard and where each of these are located. So in here instead, I would actually give you a list of the documents, parts or components within the sixty one 850 and what are they actually in this chapter? So first of all, we got the IEC sixty one 850 Dash one, which is the introduction and overview of the standard. Then we will have to Dash two, which is the glossary. The Dash three is the system and project management. Dash five is the communication requirements for function and device models, and Dash six is the configuration and language for communication in the electrical substations related to IEDs.

Now then, the Dash seven is actually quite a big part because they have the Dash one two three four Dash four one zero four two zero and five one zero. So the seven Dash one, well, first of all, the chapter seven or the Dash 7s are all basically the basic communication structure, with seven Dash ones talking about the principles and models. Seven Dash two talking about the abstract communication service interface or the EXI, the seven Dash three being the common data classes and the seven Dash four are the compatible logical no classes and data classes. Now the four, 10, 420 and the 510, these are basically the applications. So the 410 and the seven Dash 410 are hydroelectric power plants.

Communication for monitoring and controlling the seven Dash 420 is the basic communication structure for the distributed energy resources. Logical nodes and the seven Dash five 10 is the basic communication structure for hydroelectric power plants in terms of the model and concepts and guidelines. Now, moving on from the Dash seven chapters, we have the Dash eight, where the Dash eight Dash one is the specific communication service mapping or the CSM mappings to manufacture messages specification and S and to ISO IEC. Eighty eight 02 Dash three with respect to the chapter nine or the Dash nine portion, you got Dash two and a Dash three with the Dash to being specific communication service mapping sample values over eight eight two Dash three and the Dash nine Dash three is the precision time protocol profile for power utility automation.

Then you will have chapter 10, which is the conformance testing. So after chapter 10, we'll jump all the way to 80 in which we have Dash one, Dash three and Dash four. So eight. Dash one is the guidelines for exchanging information from a CDCC based data model using 68 70 Dash five one two one or sixty eight seventy Dash five, Dash one or four eight three is the mapping to where protocols, which is the requirements and technical choices, and 80 Dash four is their translation from the C O S M object model to the sixty one 50 data model. Then the last chapter is the Dash 90, which has a lot of chapters from one two three four five seven eight and then 12. So 90 Dash one is the use of the IEC. Sixty one eight fifty four.

The communication between substations chapter 90 Dash two is using sixty one eight fifty for communication between substation and control centers. 90 Dash three is using sixty one fifty four condition monitoring, diagnosis and analysis. 90 Dash four is a network engineering guideline. 90 Dash nine is the use of IEC sixty eight fifty to transmit synchro phaser information regarding to C thirty seven point one one eight nine. Dash seven is the object model for power converters in distributed energy resources, or ADR systems. 90 Dash eight is the object model for e-mobility and. 90 Dash, 12 is the wide area network engineering guidelines.

Now, as I said, the IEC sixty one age 50 is a very, very meaty standard. And as such, I'm not going to go over the information for all chapters of the standard, nor with any professional engineers who are actually required to know the in and out of every single document within the IEC. Sixty one eight fifty. But it is highly recommended to read it through and to know where everything is just so that when it needs be, you could actually quickly find a reference for the information that you need.

The Data Model

IEC sixty one eight fifty is the most promising standard for modern utility grids when it comes to interoperability. The most notable advantages of IEC sixty one point fifty over other common standards are self describing devices and object oriented peer to peer data exchange capabilities. Among the various essential features and benefits are the use of names for all data virtualized models, standardized configuration, language, reduced cabling and lower transducer transducer installation costs. The IEC sixty one age 50 object oriented virtual model, which comprises standardized data and data characteristics, is made up of what we called the logical notes, which are abstract data objects.

The EXI is defined by IEC sixty one 850 and it generates objects and services that are independent of protocols. This provides a hierarchical class model in which a communication network may access all class information services that act on these classes and its related parameters on an Ethernet data frame. The abstract interface allows data objects to be translated to any other protocol, such as the mass and the small particles. The physical or logical device and a number of logical nodes are represented by the virtual model. The International Electoral Technical Committee, or IEC six 20 50 classified logical nodes into logical categories. Each logical node has data which are standard and linked to the function of logical nodes.

CDC or Common Data Class, makes up the majority of data objects, which include fundamental data items, data control and measurements. Each data element has a number of data characteristics, with each of which has a data type of FC or what are known as functional constraints. A logical device is an object that allows you to organize logical nodes, which are basically a collection of functions. Each logical device must have a logical node that holds nameplate information, which is represented by the label AL and zero, and it must be present within the device and it helps identify it to land zero definitions IEC sixty one 850 and instantiation is L and zero.

In most cases, the physical device health is also present in a logical device and is represented by an instance of LP HD. Other logical nodes in a logical device must exist in addition to LN zero and Ph.D. in order for it to be usable for the sixty one 850 applications. These astrological notes are usually from different logical no class groups. Then the overall group, for example l x x x. A general logical note is usually added to many logical devices to reflect the physical device, name, plate and health information. Logical devices can be arranged in a hierarchy with the parent logical devices and zero controlling the mode and behavior of the hierarchy's offspring.

You can also change an existing present name or instance of logical devices, the Con F L D name server as your capability controls the ability to alter a logical device's preset name or instance if set to true throughout the IEC. Sixty one eight 50 engineering processes, the values of that initial supply instance and configured name can be redefined. The name of a logical device that results thus serves as the foundation for all object references within that logical device. Within an IED, there is an explicit technique for expressing the hierarchy of logical devices.

The hierarchy is utilized to define the management hierarchy of specific functions.

The hierarchy of functions and sub functions for logical nodes inside a logical device was used to build the logical device management structure, except for a logical node that monitors a device physical health, which is too low tech that I said earlier. All logical nodes can have their mode and therefore their behavior changed by menu plating the mod data object. This implies that the mode of another function, such as a measurement function, can be changed by interacting directly with the logical node. All logical note modes inside a logical device, on the other hand, may be controlled by altering the mode of the land zero included within that specific logical device. So now let us look at the diagram presented in front of you right now.

The management hierarchy is based on a logical node design pattern. The logical device hierarchy is a list of Elon zeros that have been instantiated in various logical devices of value. In one line zero set in the name of the logical device that includes the mastering L and Zero is used to indicate the hierarchy. In this case, Elon zero in Logical Device one controls the L and zero of the logical device to the logical device hierarchy created by the data object, a graph in reference to the L and zero logical node. The process for changing mod information and a logical device hierarchy within the same IED is a local problem that is not susceptible to sterilization, although it usually involves some sort of internal messaging or data sharing.

The IEC six 20 50 standard also allows for hierarchy of ideas to govern the behavior of other items in the system. For example, a b control is often used to coordinate the behavior of ideas with any substation, be the logical device hierarchies were used to enable for the creation of a hierarchy based on RF values. Specifying an L and zero in a separate IED exchange method must be set up since the mod exchange from the controlling logical device must be transmitted across the network. The logical node TRF construct is commonly used to do this. The logical device with no TRF value must have an instance of the physical health functions within an internal IED hierarchy. For example, the Ph.D.

note, other logical devices may also contain instances of that Ph.D. unless a logical device is serving as a proxy for another logical device. Logical nodes are functions that are used to automate, monitor and create distributed collections as the number and kinds of logical node definitions for the 650 grow. So does the number and types of application areas. Substations, wind power, hydropower and the distributed energy resources are now among the application domains within the scope of IEC. Sixty one 850 outside of substations and distribution functions. The logical no cost or the line costing is assigned a character that symbolizes the domain that is responsible for the specification.

Additionally, all logical nodes have a namespace you are right or a uniform resource identifier that indicates the domain from which they're defined and governed. Traditionally, all logical node definitions were inherited from logical nodes with the addition of UML modeling. There's no hierarchy of abstract logic, no classes from which all logical node definitions are derived. The drawbacks of using the inheritance hierarchy is that there is no way to determine the order of the data objects within the loincloth's declaration with perfect certainty. SQL or substation configuration language, on the other hand, takes a different approach to logical nodes.

It is used in the engineering process to segregate logical nodes, client plans and logical nodes contained by survey and logical devices as the basis for the separation. A logical device can include any logical node and GIF information that can be shared among various ideas. Client plans are used to set up subscriptions such as reporting goose and sample values, but they don't expose any data that may be shared. Client plans are generally specializations of the abstract logical node, a non-process interface alone. These are usually designated with an L x x x, although they can be any logical node that is not exposing information to be shared. Reporting goose and sample values are all used by client lands.

Client lines may also issue control commands and write data, as well as set set points to set in groups and do other task client lance or specified by their location in the ACL found not by their logical node hierarchy. So the Client L ends or the non processing interface L ends. So these are just the client logical nodes and in non processing interface, logical nodes. The IEC sixty one 850 specifies a language and representation format for the configuration of electrical substation equipment in the form of ACL or substation configuration language, which is formally known as substation configuration description language. The IEC 620 50 Dash six standard specifies the entire ACL representation and its features, so be sure to read it up if you're interested in the subject.

It comprises data representation for substation device entities, as well as logical nodes, communication systems and capabilities for the related operations. The ability for various devices in a substation to exchange files and have complete interoperability is enhanced by the comprehensive representation of data within the ACL. A typical ACL file contains five parts and they are the header substation communication IED, as well as the data type templates. The header, which is the first chapter of the ACL configuration file, is used to identify diversion and other basic information. The substation chapter deals with many components of a substation like as devices, interconnections as well as other functions, power transformers, voltage levels, bays, general equipment and conducting equipment such as breakers are among the elements.

Logical nodes that describe functionality linked to the item and in a substation are referred to the substation from the substation component. The communication chapter covers the many communication points or access points for gaining access to the various IEDs across the system. The chapter includes several such networks and access points and intelligent electronic devices, and set up as described in the IED chapter, and it contains various IED access points, logical devices, logical notes, report control blocks and other items that fall under the IED. It specifies what data and IED publishes as reports and a generic substation events, as well as what gauss or guest data it is set to accept from other IEDs.

Lastly, the data type template defines various logical devices, logical nodes, data and other features that are divided into several instances. This chapter of ACL depicts entire data modeling, according to 61 850 Dash seven, Dash three and seven Dash four, and it has broken down even further into the L now tied the D o tied to a type and the end NewME or the enumeration type. As for the types of school files, they're generally six types, and they are the IED capability description file, which is the ICD, the configured IED description file known as the CID to instantiate it IED description file, which is the IED, the systems specification description file, which is the SSD, the substation configuration description file, which is the seed and the system exchange description file, which is the s e d.

First of all, the manufacturer provides the ICD file, which specifies an IED as the whole capability range. The file comprises a single IED chapter, with communication and substations chapters available as options. The C.I.D files are AC Rd files reduced down to the essential parts for the IED. Linking the IED configuration tool and the IED, it contains everything the IED requires from the CD Files two to be set up, such as the data definition behind a goose message the IED will receive. The IED file specifies an IED configuration for a given project and serves as a data transfer format between the IED configurator and the system configurator. It has an IED chapter, a communication chapter with parameters, data type templates for the IED and optionally a substation chapter with function or logical notes connecting to the single line diagram.

The SSD file provides the full substation automation system specification, including the substation, single line diagram and capabilities, or logical nodes. The SSD requires the substation portion, data, type, template and logic. No types to be defined, but the IED chapter is not required and it's optional. The C d file contains chapters on communication, IED and data type templates explaining a complete substation in full as a result, city files compliance ICD and SSD data. Lastly, the S e d file is a subset of a C default that includes extra engineering rights for each IED, as well as ownership of all ACL data. It is used to share system settings between projects defining how one object's API may be utilized by another.

The file format for SQL is experimental now XML stands for Extensible Markup Language, a markup language is a collection of codes sometimes known as tags that define the text in a digital document. Hypertext Markup Language, or HTML, is the most well-known markup language, and it is used to format web pages. Example, a more versatile relative of HTML enables complicated data to be conducted over the internet or network. In this case, a simple example of what it will look like is what is shown in front of you right now. Now, don't worry about the raw format of these SQL files. That's that much as there are designer tools to generate these files to create the substation configuration description or c d for the substation design.

The engineer usually imports the ICD files of each device into the system configuration tool. The capabilities of all devices necessary for the planned system are included in the CD. The city's instantiated IED description, or the I.D. file, may be exported using the designer tool and in terms of function and objects. The idea is a subset of the ICD, but it also includes configuration information relevant to the system design, such as reports and goose subscriptions. Now, an IED configuration tool may produce a configured IED description or the CID file containing the identified specific information, which could include mapping information to instruct the device how to configure itself and as a result, when employing IED imports.

I think IED engineers generally utilize a designer tool to create these C.I.D files. So as I said earlier, don't worry about the XML format that much because usually you don't really need to know it in detail. Logical nodes are classified into functional groups, of which each group contains multiple data classes. Now I will not go through each one of them, as you can do, so once you get your hands on the standard. But here I will go through some of the functional groups just to give you an example. So let us start off with a system. This functional group is where you have information on the IED template or the and zero health or the Ph.D., as well as supervision for exchange methods, including goose sample values, reporting and telling control communication channel information. Now next is the interfacing and archiving.

This group represents a collection of functions and generally represents client functions for reporting groups and sample values that are utilized for subscription reasons. The I am my function, which is something they have provided as an example before in the L format chapter, is commonly used to represent the operator interface of a skater system or a substation. For example, its user interface. Note that these functionalities don't always need the use of an IEC sixty one 850 server in an assault. These functions are virtually never restricted to primary or secondary equipment. The Metering and Measurement

Functional Group represents a set of functions for data acquisition of different values in both AC and DC systems, such as power related information.

For example, amps voltages, frequency, etc. Now the meteorological met or environmental or envy, as well as the hydro, the hydrological or the M H Y D functions, are generic functions that can be used in any of the indicated areas. Then there is actually the scheduled function, which is the FCC and the ability to control which schedule is performed, and these are all general functionality represented by this collection of methods. So the generic function is basically for tags of or stuff that is very much not associated with any other groups. So it's miscellaneous. The last one that I want to talk about is the control function, which represents a collection of functions for controlling and interlocking switches. These functions are usually used in conjunction with logical nodes that reflect the functioning of the primary or secondary equipment under control.

Now, as I said before, and I will say it again, this is not a comprehensive list, so I have just gone over some examples. The IEC 650 standard actually contains a table of all of the functional groups and the stuff that belongs within those groups, so be sure to check it out to get a reference of what you need in your system. Now, the notion of common data costs, or CDC, was created to identify common building blocks for constructing bigger data objects. The ability to reuse status, control, setting and description definitions is what is the main use for the common data costs or to CBDCs. Now there are many CBDCs, and they're usually grouped into categories of status, measure, control and the setting and descriptions of each class.

For example, the single point the IS class under the status category contains a set of attributes such as the festival that stores particular data for that class, which in this case, it's for the festival. It stores a Boolean value for the single point status. Now there is one property of an attribute that I would like to bring attention to, and that is the functional constraint. The functional constraint, or AFC, is a detachable characteristic that describes how the attribute is used to give structure and context. It is useful to functionally arrange data characteristics, so some of the functional characteristics are what is shown in front of you right now, which are, for example, AM X equals two measurement information. S.T. is the status information as P is setting SB substitution. C F is configuration.

D C is a description. S.G. is the setting group s e is the setting group. Editable SRS Service Response O R is operator received. BLE is blocking, X is extended definition. C o is control. Now these are only some of them. There's actually a much more comprehensive list, so if you want to see a more comprehensive list, be sure to check out the actual IEC sixty one eight 50 standard under the document seven Dash two specifically. So now let me describe the overall modeling approach as a conclusion to this chapter. Now a physical device is the starting point for the IEC sixty one to fifty device model. The device to connect to a network is what we refer to as a physical device. The network address of a physical device is usually used to identify one or more logical devices that may exist within each physical device.

And the IEC 650 50 logical device model allows a single physical device to operate as a proxy or gateway for many devices, resulting in a common data concentrate representation. There are one or more logical nodes in each logical device. A logical node is a designated collection of data and services that is logically connected to a power system function as a suffix to the logical no name. Each logical note has an L and an instance ID. Ecological notes may additionally include an optional application specific L and prefix to better identify the logical notes function. Each logical note has one or more data items. Each data piece has its own name. The standard determines certain data names, which are functionally connected to the power system's purpose.

Each data element within the logical node complies with the IEC. Sixty one 850 Dash seven Dash three definitions for common data class, of which I have described earlier. And each of this common data class, or CDC, identifies the data type and structure within the logical node. CBDCs exist for status information, measured information, controlled status information, controllable analog set point information, status settings and analog settings, just to name a few. Each CDC has a unique name, as well as a collection of CBDC characteristics and each having its own unique name, type and function. Each CDC attribute is then part of a set of functional constraints, or FC, that organizes the characteristics into categories. The IEC six one model of a device is a virtualized model specified in IEC.

Sixty one fifty parts seven. That's sort of an abstract representation of the device and its object. Then, as outlined in the IEC six 20 50 Dash eight Dash one. This abstract model is translated to a particular proper protocol stack based on the MSA, which is the ISO ninety five 06 standard, the TCP IP and the Ethernet layer. The IEC six 20 50 Dash eight Dash one provides a technique of converting model information into a named mass variable object that results in a unique and clear reference for each element of data in the model. When mapping the IEC, sixty one to fifty objects to AMS.

Generic Substation Events

As specified by IEC, sixty one 850 generic substation events, or GST, is a control paradigm that provides a quick and reliable mechanism for transmitting event data throughout whole electric substation networks. Multicast or broadcast services are used in this paradigm to guarantee that many physical devices get the same event message. Now, this is actually further split into two big categories called juice or generic object oriented substation events and GST, which is called the general substation state events. Goose is a standardized model method in which any type of data which can be status or value is combined into a data set and delivered within a time period of four milliseconds. A number of methods are employed in order to achieve the desired transmission, speed and reliability.

Now that generic substation state events or GST, on the other hand, is an expansion of the event transfer that is set out in the U.S.A. 2.0 or what we call the utility communications architecture. Now, as with Goose, GSC only allows the exchange of status data, which is represented by a status list, which, on the other hand, is just a string of bits rather than a whole data set in general, GSC format is simpler than goose, which means that it is also processed quicker over time. However, the use of Goose is gradually replacing GSC, and thus the support for GSI will eventually be totally phased away. So since GST is phasing out, let's talk about Goose. Let's focus our explanation on a goose. The Goose Protocol is a protocol that's based on events.

The goose communication idea is that the publisher delivers messages on a regular basis, and when an event occurs, such as a relay trip, it sends a burst of messages with fresh data because the protocol is built on a publisher subscriber relationship. There is no guarantee that the message transmitted is successfully received by the subscriber and therefore the message first reduces the risk of message loss. Our system wide distribution of input and output values is possible using the generic object oriented substation event model. By using Ethernet multicast, a device may broadcast goose messages to a large number of recipients. At the same time, data sets are what is used to create these messages, while 61 eight 50 enables any data set to be used with goose.

The actual applications indicate that the dataset should only contain a few status values and their associated quality information. A change in the value of any dataset member is deemed a change of state, and the new data is then instantly published in the event of loss packets or new devices coming online that require the present status. The messages were sent to save bandwidth at the duration between retransmission increases, with time from meantime immediately after state change to max time in steady state, goose follows what we call the publisher subscriber structure. For example, substation devices that need a particular device's output or a function from another device expressed an interest in the data and therefore subscribes to it.

The network distributes the specific data after the publisher releases it. The publisher, like people on Twitter or YouTube making tweets or chapters, does not need to know how many subscribers there are, nor who they are. This allows the publisher to accommodate an endless number of subscribers using a scalable infrastructure. The publisher and subscriber situation that I've just mentioned is actually enabled by the concept of multicast. Multicast is a type of computer networking communication in which data is sent to a number of target machines. At the same time, multicast distribution can be one too many or many too many.

Multicast should not be confused with point to multipoint communication at the physical layer, though multicast networking is built on the basic idea that a server may send a single packet to a large number of recipients. Multicast differs from broadcast in that it is more targeted. Multicast packets are only received by receivers that request them. Unlike broadcast packets which are received by all receivers in a network segment or broadcast domain, multicast receivers can also be placed behind routers across a wider network. Multicast has a few key properties that determine how it is being utilized.

It is beneficial in cases when a large number of receivers need to receive the same data, since a server only needs to send each packet once and it will reach all of them because the network, rather than a head and server, replicates and distributes these packets. It skills effectively to extraordinarily high numbers of recipients. However, because the multicast is one way. Any replies would require a different protocol. This also implies that dropped packets must be insignificant, or that data recovery techniques must be designed independently. Now, back to the topic of goose. The communication network uses multicast addressing to execute, goose, publish and subscribe.

Topic routing and content filtering is dependent on the implementation, receiving the message that was routed to it based on a subject. Producers can publish to the same topic or multicast address, have it delivered to the subscribing implementation and have the subscriber decide on the necessary processing of that information. The content filtering is based on several items that are embedded into the goose message, and they are the control block named the data set. The config revision and the App I.D. Now to control block information is encoded as an object reference and a goose message which contains the control box complete path reference. The name of the logical device is included in this reference.

Filtering based on this data enables for the distinction of message sources that use the same topic along with the control block. The dataset information is included as an object reference in a goose message, along with the data sets. Complete Path reference the name of the logical device, as well as the logical nodes, are included in this reference. The actual information be provided in the message as defined by the dataset reference name filtering on this data and tables for the distinction of message sources that use the same topic, as well as assessing whether the information supplied is what is expected. Now, as for the config revision, it is included in the goosed message and helps the subscriber to identify if the datasets contain a change from what was expected.

Lastly, the app idea is included in the goose message, allowing the subscriber to filter the list that delivered the message based on a distributed application in which the message is involved. Lastly, I just wanted to briefly talk about sampled values. Sample values, or S.V., is a technique for publishing sampled and analog data from measuring equipment such as current and voltage transformers. A device publishes sampled values messages through Ethernet multicast so that they can be subscribed to by any number of other devices. Unlike goose, sampled values are sample driven rather than event driven. The samples are taken at a certain sampling rate and with a particular data set in mind. Publishers of sampled values are commonly referred to as emerging units, and subscribers are a form of IEDs.

SNTP

In this chapter, I want to have a quick talk about S.P. Now, S.A. S.P. stands for Simple Network Time Protocol. So let me start with talking about A.P. first, which is simply just network time protocol. The Network Time Protocol, or NPP, is a networking protocol that allows computer systems to synchronize their clocks over packet switched variable latency data networks and TCP is one of the earliest internet protocols still in use. Having been in use since before 1985, actually now A.P. is designed to keep all participating computers in sync with the coordinated Universal Time or UTC, and it's going to be as accurate as to be within a few milliseconds.

It selects correct time servers using the interchapter method, a modified version of what we call it, the more zulu's technique, and is meant to reduce the impacts of changing network latency under ideal conditions. A.P. can keep time to be within tens of milliseconds across the public internet and better than one millisecond accuracy in local area networks. Asymmetric routes and network congestion might result in delays of up to 100 milliseconds, though the protocol is commonly defined in terms of a client server architecture. But it may also be utilized in peer to peer interactions, where both peers regard the other as a potential time.

Source implementations use the user data and protocol or the UDP on Port one two three to send and receive timestamps after a first round trip calibrated exchange. They can utilize a broadcasting or multicast in which clients passively listen to time updates, and TP sends out a notice when a leap second adjustment is come up with. But no information regarding local time zones or daylight saving time will be sent. Now, A.P. employs a semi-layered hierarchy, hierarchical set of time sources; each stratum in this hierarchy is given a number, starting with zero for the reference clock at the top.

A server running at Stratton and +1 is sent to a server. This number is used to prevent cyclical dependencies in the hierarchy by representing the distance from the reference clock. It is normal to discover three times sources are greater quality than other strutton two time sources. Now, Stratham is not necessarily an indication of quality or dependability. However, there are 17 strands starting from scratch zero where the last one, stratum 16, is an indication that the devices are synchronized. So let us not look at the first three stratus. The atomic clocks, GPS and other radio clocks are examples of high precision timekeeping technology.

That is what we call the Stratos Zero. They send highly precise pulse per second signals to a connected computer, which causes an interrupt and a timestamp to be generated. The Stratham one, on the other hand, is a computer whose system time is synced to the connected stratum zero devices to within a few microseconds for sanity checks and backup. Stratham one servers may pair with other Stratham one servers. Primary time servers are another name for these Stratum one servers. Now lastly, Stratham two are machines that are linked to Stratum one servers to a network. Frequently, Ashram to machine will query several Stratum one servers. Peering between SRAM, two computers can give a more consistent and resilient time for all devices in this peer group.

And as you can see, like afterwards, you can actually see the pattern between Straten. One is from two, and that actually continues on up to Stream 16. So that's why I'm just going to stop in Stratham too. A.P. clients and servers are synchronized to the coordinated Universal Time, or UTC time frame, which is used by national laboratories and broadcast to radio, satellite and phone modem. This is a global timeframe that is unaffected by geological location. Now there are no provisions for local time zones or daylight saving time. Correction. Nevertheless, the operating system can execute these operations on a per user basis. The rotation of the Earth about its axis determines the time frame on which UTC is based.

A leap second is introduced at intervals of roughly 18 months, as calculated by the International Earth Rotation Service to utilize UTC in relation to take my. So what is the difference between a SMTP end and deep and A.P., while a full featured A.P. server or client achieves a high level of accuracy and avoid abrupt time steps as much as possible by employing various mathematical and statistical methods as well as smooth clock speed adjustments, SMTP is only suitable for simple applications with low accuracy and reliability requirements when compared to a complete A.P. implementation.

Essence achieves just a poor quality time synchronization by neglecting drift values and utilizing simplified system clock adjustment methods with typically just simple time stepping RC 20 30 defines centipede version. For us to be utilized exclusively at the extremities of the synchronization. Subnet S&P clients should only be used at the next highest strand and in settings where neither A.P. or S&P client is synchronized by another S&P client. S.A. servers should only be used at the subnets Root or Straten one. And in situations where there is no other source of synchronization available, such as a trustworthy radio or mobile modem time service, only a comprehensive implementation with redundant sources different from routes and well-developed algorithms can provide the level of dependability often anticipated of mean servers.

Now, as you can see, A.P. and S&P revolve around the topic of time. So how is this being used in IEC 61, eight 50? Well, generally this is actually used for timestamps. You might have guessed, I assume the timestamp is in UTC time rather than in local time. The Leap Second information is one of the most contentious aspects of the timestamp. This is important because if a leap second happens, it is conceivable for two separate events to occur at the same timestamp. Since midnight repeats one, a leap second is applied. As a result, forensic and some real time analysis are quite challenging. The International Astronomical Society has linked the leap second to the Earth's rotational speed, and several requests have been filed to end leap seconds.

But the astronomical society has so far rejected the timepiece that does not use leap. Seconds is referred to as to why the problem with guys is that it makes it considerably more difficult to convert it to local time and needs maintenance every time. A leap second occurs until leap seconds are removed. Time synchronization is necessary to produce correct timestamps. Network Time Protocol and P2P are the two forms of time.

Synchronization is officially supported by IEC sixty one point fifty. Now I've described it in TPY earlier. So now let us look at P2P, which stands for Process Precision Time Protocol in P2P. The network infrastructure calculates the time offset in real time as the time messages go through the network's different hops, rather than the entity doing so.

A grandmaster is the originator of the initial time synchronization message. It is usually thought of with the help of GPS. This sends out a sync message of the original time, plus an offset of only two packets delay in reaching the network when a sync message enters an Ethernet switch. The packet must be processed as a transparent clock. The delay between ingress and egress must then be calculated, and the offset inside the sync packet must be updated. When the end device gets the sink message, it must calculate the ingress processing delay and have the following information available to compute a timestamp to compute its new time.

It uses the grandmaster or original time plus network offset in the packet plus ingress delay. In most cases, the calculation allows for temporal precision of less than one millisecond. He also has the benefit of being able to have many grandmasters on a network with a specified process for picking the best master. As a result, he has built in redundancy and fell over capabilities for IEC six hundred fifty applications. Clock synchronization and accuracy are critical for client, server and goose time stamping, and TP synchronization is good for CPTPP teams, of which I mean like these are sample values and synchrophasor applications. However, P2P or RB or GG should be utilized instead of S&P. The logical node LTM, as is provided by IEC sixty point fifty to oversee the kind and source of synchronization for the different types of data that you have been collecting and transmitting throughout. Utilizing the one 850 standard.

MMS

In this chapter, I'm going to talk about the EMS component of IEC sixty one, eight 50. M.S. or manufacturing message specification is a globally standardized messaging system for exchanging real time data and supervisory control information between network devices and or computer applications that is dependent on one. The application function being performed or to the device or applications developer. Now, MMS is an international standard, which is from the ISO ninety five or six established and maintained by the International Organization for Standardization Technical Committee number one eight for which it's also codenamed TC one eight four. Now, MMS provides messaging capabilities that are generally general enough to work with in a wide range of devices, applications and industries.

The MMS services, for example, allows one app or device to read a variable from another app or device. The services and messages are the same whether the devices have the posse or programmable logic controller, or even a robot in sectors as disparate as automotive or electrical utility and space exploration and MMS has been put to use in applications as different as material handling, fault Annunciation Energy Management, electrical power distribution, control, inventory control, as well as deep space antenna locations. Now, a mess offers a diverse collection of services for real time peer to peer communication across a network. Many typical industrial control devices, such as science postseason robots, have utilized MMS as a communication protocol.

Remote terminal units, energy management systems and other intelligent electronic devices like reclosers and switches are examples of EMS uses in the electrical utility business. MMS connections are accessible on most common computing systems, either directly from the computer manufacturer or through a third party. MMS is especially well-suited to any application that necessitates a common communications method for a variety of communication tasks linked to real time access and dissemination of process data, as well as supervisory control. MMS is rigorous enough to reduce disparities between applications that perform similar or complementary activities, while being general enough to be used in a wide variety of applications and devices.

Insufficiently specified communication methods might result in applications that execute identical or commentary are complementary activities in different ways. As a result of the developers differing implementation decisions, the apps are unable to interact with one another. While conventional communication methods merely allow a series of bytes or a message to be sent across a network, MMS goes much beyond. This is because MMS also gives messages, clarity, structure and meaning, making it far more likely for two separately created applications to work together. MMS offers a collection of capabilities that make real time data distribution and supervisory control tasks by a network in a client server context as easy or complex as the application requires.

The virtual manufacturing device, or VMD concept, is a fundamental element of MMS. The VMT model describes how MMS devices, also known as servers, operate when viewed through the eyes of an external MMS client program. MMS enables any application or device to serve as both a client and a server at the same time. The VMT model gives client applications a consistent and well-defined view of the items in the VMD. MMS services are used by clients to access and menu items. All servers must follow the VMD model, according to MMS. The VMD model defines the following in general, which is the first one, which is the objects or, for example, variables. The second one being the client site services for accessing a new plating. These objects, such as read it and write these variables.

And third, the server's reaction to receiving these service requests from the clients. The VMT may be thought of as a master object over which all other mass objects are subordinate to, and that is because verbose domains, etc. are contained within the VMD for getting information and status about the VMD mass provide services such as status, unsolicited, unsolicited status and identify option. It also includes functions such as get, name, list and rename for maintaining and retrieving information about VMT objects. MMS is a versatile and comprehensive framework for transmitting changeable data across a network named or unnamed, which is addressed and named.

List of variables are also supported by the MMS variable across mechanisms. The type description of the variables can likewise be modified as a distinct mass object in the MMS and MMS, variables can be basic. For example, it can be an integer Boolean floating point or string, or it can be as complicated as having arrays or actual structures with it. The client server connection between networked applications and or devices is an important component of the VMD paradigm of AMD, and its objects are stored on a server, which is a device or program, a client as a network program or device that asks the server for data or performs an action. A client in this broadest definition is a network entity that sends MS Service requests to a server.

A server is a network node that replies to the client's MS queries, while MMS provides services for both clients and servers. The VMT model only defines server behavior that is observable on the network. MS client and server features are available in many MMS apps and compatible devices. Only the server functionalities of such apps will be defined by the VMD model for all network physical features of the server application device. Every MMS application or device that offers MMS service server services must follow the VMT paradigm. Clients of MMS are only obliged to follow the standards that govern message structure, construction and sequencing.

The VMT model includes a configurable execution model that specifies how the mass server's program execution may be managed. The definition of the domain and program invocation objects are at the heart of this execution paradigm. The mass domain is a named mass object that represents a resource on the physical device. This resource can be anything that can be represented as a single and Typekit block of data, which we refer to as low data. Domains are used to represent regions of memory and a device in many common applications. The latter program memory of HPLC, for example, is generally represented as a domain. The one thing to keep in mind is that some programs enable you to express blocks of variable data as both domains and variables.

MMS does not define active domain content and doesn't place any restrictions on it. This would be the same as declaring a real object. For example, the latter program, the VMT implementer, is in charge of the domain's content. An event or an alert is a simple to define in the actual world within their particular field of expertise. Most people have an instinctive sense of what can make up an event in a process control application, for example. It is typical for a control system to raise an alert when a process variable, for example, temperature pressure exceeds a predefined limit known as the high alarm threshold. For example, when the phase angle difference between a current and voltage waveforms of a power line exceeds a particular number of degrees, an alert may be produced in a power distribution application.

The mass event management paradigm provides a framework for assessing and controlling these types of events. Network connection elements. This is achieved by establishing three named objects that each represent a different type of data, the first one being the current state of a situation in which we'll call it the event condition. The second being who should be notified when an event occurs, which is event enrollment? And lastly, the Book of action that the VMT should take in the case of an occurrence which we call it the event action. The mass event management paradigm should be utilized when the application is more sophisticated and requires a more thorough description of the event environment in order to assure compatibility.

A named object that reflects the current status of some actual condition within the EMD is what we called an MMS event condition object. It is worth noting that MMS doesn't specify the VMD action that causes the event condition to change status. The mapping between the high alarm limit and the status of the event condition is not clearly defined in MMS. And AMC does not provide the necessary setup or programming to construct a mapping between a high alarm limit and the states of the event condition. Even if the high alarm limit is represented by an EMS variable, the change in the status of the event condition, according to the MMS, is triggered by some autonomous activity on the part of the VMD that is not described by MMS.

When the status of an event condition changes an event action is called a mass object that indicates the action that the VMD will perform. Now it is not necessary to do an event action. The VMD would perform its no event notification process for processing and event action if this parameter was just left blank. Now, an event action is always characterized as a confirmed EMS service request when it is used when an event enrollment is specified. The event action is associated or linked to an event condition. A technique through which an application can regulate access to a system resource is required in many real time systems. A workstation that is physically accessible to many robots is one example.

It is necessary to have some way of controlling which robot or robots have access to the workspace for these purposes. MMS defines two types of semaphore. One is the token semaphore and the other is the poor semaphore. As a token semaphore as a named MMS object that represents a resource under the control of the VMD and to which access must be restricted. Using MMS service as a token, Semaphore is depicted as a collection of tokens that MMS clients take and release control over. The

Semaphore might be owned by several users or can be only one user as well. Now, when an MMS client processes the token, it has access to the underlying resource to some extent.

To some extent users, for example, could both wish to modify the set point for the same thing at the same time and to coordinate their access to the set point. These users might utilize an MMS token semaphore to hold only one token to indicate the object of being targeted. When a person owns a token, they have the ability to alter the set point. The other would have to wait until the owner, the first user, relinquishes control in order to make any changes in your turn. Now, a token semaphore may likewise be used to coordinate the operations of two MMS clients without having to represent any real resource.

This type of virtual token semaphore has the same appearance and behavior as the real thing, except that MMS clients can build and remove them using something called the defined semaphore service. A pool semaphore, on the other hand, is almost identical to a token semaphore, except that the tokens are individually identifiable and are also given a name when making tick control requests to MMS content optionally specify these named tokens. The pool semaphore is an MMS object in and of itself. MMS objects are not contained in a poor semaphore as name tokens. The poor semaphore, his name is distinct from the names of specified tokens, and only genuine resources within the VMD can be represented by poor semaphore. And as a result, Poor Sam, of course, can't be generated or removed with mass service requests.

Ethernet

Ethernet is a group of wired computer networking protocols that are frequently used in local metropolitan and wide area networks, or you probably more commonly heard as the land's man's and the lands. It was originally commercially available in the 1980s and was standardized in the name of AI Tripoli eight zero two point three back in 1983. Ethernet has subsequently been improved, enable faster data rates, a larger number of nodes and longer network links, while still maintaining a high level of backward compatibility. Its ability to adapt and offer better levels of performance while preserving backward compatibility assured its continued appeal as network technology and vault evolved in the late 1990s.

Ethan, its original 10 megabits per second throughput, was raised tenfold to 100 megabits per second, and I hope we continue to improve the speed with subsequent upgrades. Current Ethernet variants can handle data rates of up to 400 gigabits per second. Now, the stream of data is divided into smaller bits called frames by systems interacting over Ethernet. Each frame comprises source and destination addresses, as well as error checking data in order to detect and reject damaged frames. In most cases, higher level protocols initiate retransmission of lost frames. Ethernet offers services up to and including the data connection layer, according to the Osai model.

Other actually A02 networking protocols, including actually able two point one one Wi-Fi and FTD, adopted the forty eight bit Mac address. No such network access protocol Snap Headers also utilized Ethernet Ether type values. Packets and frames are the two types of transmission units defined by Ethernet. The payload of data being ProCase broadcast is included in the frame, as well as the standards and the receiver's physical Mac addresses. The information on the virtual LAN tagging and quality of services. And the information on error correction to detect transmission issues. Now, each frame is enclosed in a packet that comprises several bytes of data that establishes the connection and marks the start of the frame.

A unique identification issue to a network interface controller for use as a network address in communications inside a network segment is known as a media access control address or commonly we call it the Mac address. Most assuredly, eight zero two networking technologies such as Ethernet, Wi-Fi and Bluetooth employ this technique. Mac addresses are utilizing Linklater's Medium Access Control Protocol, a sub-layer of the open systems interconnection or the Osai Network model. The Mac addresses are generally written as six groups of two hexadecimal numbers separated by hyphens, colons or no separate at all because Mac addresses are mostly assigned by device makers. They are also known as burned in addresses, ethernet hardware addresses, hardware addresses or physical addresses.

Each address can be stored in hardware such as the read only memory on the card or in firmware. However, many network interfaces allow you to change your Mac address, and manufacturers' organisationally unique ID or UI is usually included in the address. The concept of two numbers of spaces based on extended unique Nephi identifiers is maintained by the way we are used to create these Mac addresses. Early Ethernet used a daisy chain or star architecture to connect numerous devices into network segments via hubs, which are layer one devices responsible for delivering network data. Now, when two devices sharing a hub try to send data at the same time, the packets may collide, causing connection issues actually created.

The carrier sends multiple access with collision detection, or the CSM may slash CD protocol to alleviate these digital traffic bottlenecks caused by collisions. This protocol allows devices to verify whether a particular line is in use before beginning new transmissions. Ethernet hubs were eventually phased out in favor of network switches. A hub can't transfer data straight from point A to point B, since it can't distinguish between points on a network segment. Instead, whatever network device transmits data over an input port, the hub is duplicated and distributed to all available output ports.

A switch, on the other hand, intelligently delivers just the traffic meant for its devices to any particular port, rather than duplicates of all communication on a network segment, increasing security and efficiency to connect to Ethernet. Interested computers and devices must have a network interface card or and I see, much like other network types. By providing a networking foundation for implementing protocols in seven levels, the open system interconnection or the Osai model specifies how different computer devices such as network interface cards, bridges and routers exchange data over a network. Control is transmitted from layer to the next, starting at the application layer.

Now, the seven layers of the OSI model are the physical data link, network, transport session, syntax and application. So now let us start off with the physical layer. This layer allows devices to send and receive data through a physical medium, such as cables, a card or another device on an electrical and mechanical level it transports to bitstream through the network. Physical layer components are found in protocols such as Ethernet and RC 230 to now within the data link layer information data packets are encoded and decoded into bits. Using transmission protocol knowledge and management errors from the physical layer of flow, control and frame synchronization are rectified here.

The Media Access Management Layer, which regulates how networked computers get access to data and transmit it, and the logical link control layer, which handles frame synchronization, flow control and error checking, are both belong in this data link layer. The network layer, which is our third layer, is in charge of constructing virtual circuits to transfer data from note to note, using switching and routing technologies, routing, forwarding, addressing internet, functioning air and congestion management and packing and packet sequencing are among the functions belong to the network layer. Information is exchanged clearly across systems in the transport layer, ensuring complete data transmission.

The transport layer also ensures proper flow management and error recovery from start to finish. Then we have the 10 connections between applications established, managed and terminated as needed within what we call the session layer to allow data exchanges between applications at both ends of a dialogue or message. Information is converted back and forth between application and network formats inside the message, regardless of encryption of formatting. This translation converts that information into data that the application layer and network can understand. And this all happens within what we call the syntax layer.

Now lastly, the application and end user processes are supported by the last layer of the application layer within this layer. User privacy is taken into account, as well as communication, partner services and limitations. This layer includes programs for file transfers, HMD, Telnet and FTP. Now, the code format and classifying Ethernet as shown in front of you, which is speed and signal and tie max cable length and then the wiring type. So for example, if we have a 10 base, five dash talk, it means 100 megabits per second base signal and type five meter as max cable length and ts for twisted pair.

And lastly, X means it utilizes duplex capable cable. The geometric arrangement of nodes and cable links in a land is known as network topology bus and star of the two most common configurations. These two topologies explain how nodes in a communication network are connected to one another. Now, a node is a network device that is active, such as an IED, a computer or even a printer. A node might also be a networking device, such as a hub, switch or router as well. A bus topology is made up of a sequence of nodes, each of which is connected to a long cable or bus. Many nodes can connect to the bus and communicate with the rest of the nodes on a cable segment. When a cable chapter breaks, the entire segment is generally rendered unusable until the break is fixed.

The 10 base two is an example of what we call bus topology. On the other hand, 10base t ethernet, a fast ethernet, employs what we call the star architecture of a central computer controlling access. In most cases, a computer is situated at one end of the segment, while the other end is terminated with a hop or switch in a central location. Because UTP is frequently used in combination with telephone cabling, the central point can be a phone closet or another place we're connecting. The UTP segment to a backbone is simple. Now, the main benefit of this sort of network is its dependability, because if one of the point segments fails, it will only affect the two nodes on that connection.

Other network users continue to function as if that chapter of the network does not exist. Now, because ethanol is a shared channel, there are guidelines for delivering data packets to minimize conflicts and ensure data integrity. When the network is ready to send packets, notes identify when it is available for the transmission. It is possible that two or more nodes in separate parts of the network will try to transfer data at the same time. A packet collision occurs when this happens. Collision avoidance is a critical component of network design and functioning. Increased collisions are frequently caused by a network of too many users.

This creates competition for network capacity and might degrade the network's performance from the user's perspective. Segmenting the network or separating it into distinct chapters that are logically connected, using a bridge or switch is one approach to reduce and overload in the network. Ethernet utilizes a technique called carrier sense multiple access collision detection to handle collisions. This is also known as the CSM slash CD when a collision is observed or when two devices attempt to transmit packages at the same time. The CMS, the CSM, a slash CD, is a form of contention protocol that describes how to respond because Ethernet allows any device to transmit messages at any time without waiting for network authorization.

Multiple devices may attempt to send messages at the same time. Design criteria for Ethernet and fast ethernet must be followed in order for them to work properly. The electrical and mechanical design features of each form of Ethernet media determine the maximum number of nodes, repeaters and segment length and network utilizing repeaters, for example, works within Ethernet. Time limitations Although electrical impulses over Ethernet medium move at almost the speed of light, the signal travels from one end of a big ethernet system to the other. A is going to be in a finite, finite length of time.

The Ethernet standard expects that a signal will travel 50 microseconds to its destination in general, so the five for three rule of repeated placement applies to Ethernet. That is. The network can only have five segments linked, and it can only utilize four repeaters, and only three of the five segments can have users, of which the other two must be into repeater lengths. If the network's architecture reaches these repeater and placement restrictions, the transmitting station will send a packet at the time guidelines are not fulfilled. This might result in packet loss and excessive risk, sending slowing network performance and causing problems for applications.

A bridge router or switch can be used to connect different networks together when situations call for longer distances or a higher number of nodes or repeaters. These devices connect two or more independent networks, allowing for the restoration of network design requirements. Switches enable network designers to create huge, well-functioning networks. The cost of bridges and switches has decreased, reducing the influence of repeater regulations on network architecture. In the overall network, each network connected by one of these devices is referred to as a distinct collision domain.

The introduction of distributed energy resources, or D.R., into the electrical grid, has had an influence on IEC sixty one eight fifty in terms of modeling and communication profiles over the communication layer IEC sixty one in fifty has its own set of encoding standards, which is expected to of the encoding rule sets described in the ISO. ITU standards are presently used in the IEC sixty one fifty. And these are the P R, which is the basic encoding rules and ER which is the XML encoding rules. Let us start with the B R tag length value or a T L V and coding is what B is mostly referred to as the encoding of bits. Strings in b r are unique. The following fields are included in the string, which is tag length, number of unused bits and value.

B r enables very blistering encoding in which unrecognized bits may distinguish between utilized and unused bits. When decoded in b r data elements are the basic units and coding is the process of representing data as a data component. Decoding is the act of reversing and coding. Each data element consists of three parts and identifies a length and a content appended. A primitive data element has a content component that does not include any extra data items. The content component of a structured data element includes one or more data components with each component encoded in octet, with the encoded data element consisting of the whole octave sequence.

On the other hand, the principles for air encoding are significantly simpler than those for PR. The reason for this is that the goal is to create an SSD that can subsequently be used to sterilize the BNF payload. The name of the attribute on the model name, for example, becomes the name of an SSD attribute in R elements. In R, the usage of implicit has no bearing visible strings. Type might have been seen on Excel D in the SSD string. However, IEC sixty one eight 50 places restrictions on that type and in general, the encoding of P and examples through OCR is easier to comprehend than the B R encoding. Now, however, though, you just need to know that these are being used. I would say a lot of times these are usually in the back end and the configuration tool will handle everything for you, so I wouldn't worry about it.

IED

An Intelligent Electronic Device is a microprocessor based controller of power system equipment such as circuit breakers, transformers and capacitor banks used in the electric power sector. IEDs collect data from sensors and power equipment and consent to order, such as stripping circuit breakers a voltage. Current or frequency abnormalities are detected or raising lower tap positions to maintain the correct voltage level and Transformers protective relaying devices. Tap Change your controller, circuit breaker controllers, capacitor bank switches, recloser controllers and voltage regulators are all examples of IEDs. A setting file is usually in charge of the operation of these devices.

One of the most time consuming tasks for a protection tester is the checking of these setting files, not digital protective relays or IEDs that use a microprocessor to perform a variety of protective control and other tasks. A typical IED make includes five to 12 protection functions, five to eight control functions that regulate different devices and auto reclose function, self-monitoring functions, communication functions, etc., etc.. IEDs are utilized as a more modern alternative to or a supplement to classic remote terminal devices in terms of setups.

IEDs, unlike the remote terminal units to use, are integrated with the devices they control and provide a standardized set of measurements and control points that are quicker to set up and need less wires. Most IEDs include a communication connector and built-in support for common communication protocols. For example, the ANP three and the topic of this Book, which is IEC sixty one to fifty, allowing them to interact with the overall skittish system or substation plc directly. They can also be linked to a substation, RTU, which serves as a gateway to the skater server. An IED is defined as an intelligent electronic device in the IEC sixty one, 850 jargon.

This basic description, however, conceals the intricacies that must be understood when utilizing the acronym IED. Within the standard, an IED must be able to communicate via a communications medium first and foremost. Even though it is digital, a computer that uses LEDs to show status would not meet this criteria because it is delivering information visually. At the heart of an IED is a capacity to share data with other devices and entities. The sole integration or exchange pattern in most non IEC sixty one 850 protocols is that of client and server. The terms master and slave are used in Mabuse's to describe client and server capabilities, respectively.

DNP three, on the other hand, uses the word master and outstation in the same manner. Both the client and the server reflect functional capabilities, and both can be found on the same device. In many cases, and RTU proxy or gateways used to describe a device that has both capabilities, devices can also have client only or server only capabilities. Information can only be exposed to other devices if the server has this capability within the framework of IEC sixty one 850. There's also publish and subscribe integration designed that may be used if it weren't for the setup techniques provided in IEC. Sixty one eight fifty six. A subscriber wouldn't need to be either a client or a server within IEC.

Sixty one fifty. There are two types of publish and subscribe exchange protocols, which is Goose and the SMB as part of an access point or server specification in SQL. The client or service functional capabilities are exposed on a per access point basis. The Access Point service capabilities may be used to describe client capabilities and to differentiate server capabilities. So, for example, one access point may support only Kyuss and another access point US might support other server services. The service capabilities defined features like a Guo's publishing and subscription support, as well as the number of publications or subscriptions that it can handle.

Devices can enable a variety of applications, depending on their functional capabilities. The majority of applications may be divided into three categories, and that is skater automation and the sink officers. Now many of these applications can have the ability to make use of electronic sensors. Let's start with the skater category's direct connection to an IED or usage of a proxy, a gateway or the most common skater application topologies. And it's worth noting that client functionality usually necessitates the use of at least one logical node. Other functionalities are available on servers in a skater system.

RTU is made up of a client and a server, with data moving back and forth between a client and server functions. The client function gathers data from other ideas. Whereas the server function proxies or repackages for consumption by another device automation applications, the second type can make use of a skate, a structure which allows clients to communicate with one or with one another via control instructions. However, the incorporation of goose high speed exchanges distinguishes DNP control applications from IEC sixty one at 50 automation applications.

Even though an RTU may supply the automation logic, proxies are seldom seldomly used with automation applications at the substation level, IEDs and automation IDs make up the automation application. As for synchro phase application to analog distribution of the current transformer or two and the voltage transformer, the Paetz was replaced by the original design of sampled values in a similar way to stepped on transformer. The measured voltage is transformed from the primary voltage to a lower secondary voltage in the. The secondary voltage will then be calibrated and conditioned to calibrate.

A secondary voltage is then delivered to the IED inputs via copper wires. Additional compensations may be necessary depending on the length of the distribution. This makes sharing the low voltage signal over long distances and in diverse conditions challenging. If inquiring minds wonder, why not just install more? Well, the answer is that the cost of a transmission level is actually quite expensive. Additionally, because they must allow electrical arcs to ground, analog pits are actually big and take up a lot of space. And that is why the stack, which is the desk on the pits, is enormous and continues to grow as the voltage increases in most IEDs nowadays.

A digital signal processing, or DSP, conducts an analog to digital conversion, scaling, conditioning and math to create the digital data that must be processed within the IED. Now, within the DSP, the sampling rate is closely regulated and any variation in the sample rate might result in mistakes. Thus, a precise time synchronization source such as RFID or GPS is critical to the accuracy of the process information. There can be one or more access points on an IED. A single communication interface is represented by each access point, only one IP address and something that works can be assigned to a single access point. A connected access point is a binding that may be found in the communication chapter of an ACL file.

A communication interface can be made up of several Ethernet connections that provide redundancy and operate as a single interface. Simply said each access point may offer a unique set of services. An access point can be either a client or a source of information in general. Depending on the sort of application for which they're being used, IEDs might be client-server or both. IEC 620 50 Dash seven Dash two contains the abstract model of a server using the IEC sixty one 850 suite of communication services server store things that are extremely viewable remotely. Many communications services are associated with specific objects in the abstract model.

Configuration exposes those configurations' communication capabilities, which are represented in service capabilities so that the system engineering process can have the information exposed in a way that can be computationally processed without requiring human intervention and without requiring an online and communicative device. The protocol used to deliver IEC sixty one to 50 services, determines how a communicative server is created. A specific communication, specific mapping or the systems provides the mapping of abstract services to a specific protocol. The mapping from abstract services and objects to one or more actual communication protocols and physical objects is the responsibility of this CSM.

Although the underlying encoding and transmission of the communication services are different, the IEC sixty one 850 Dash eight, Dash one and Dash eight, Dash two systems and the same objects and services and abstract service mapped to a virtual manufacturing device in both cases, and the logical device is an item that must be present on the server. This object allows you to group together functions known as logical nodes and logical nodes are made up of many data objects and as well as there are several data attributes in each data item, as you've known when I was talking about it in the data modeling chapter.

Functionally constrained data objects or CDs and functioning country constrained data attributes, or FTC cases are created using constraints on data objects and data attributes. Now, if CDs and their CDs can be combined into a data set, which is a list which should be a list of these CDs and FCC, FTC is now several distinct sorts of control. Blocks can refer to these data sets, and the Setting Group Control Block is a form of control block that does not use the idea of a data set where each object has its own set of services, properties and methods for configuring them. Lastly, there are two types of access point that I want to talk about, and that is the server at and the client only access point and after a clone of a server is represented by a server at the access point.

The fundamental component of server at allows you to access objects described in the referred server using an alternative host address. And this allows a single IED to have redundant communication paths, resulting in increased system robustness. Now, the server at Construct additionally allows for customized access to and publication of server objects via a separate communication interface. So now let us talk about the second type, which is the client only access point. Most IEDs have both client and server capabilities, thus finding a client only access point. More uncommon, a data front end processor, or H.M AI interface is a common exception to the norm, so those are the ones that will probably utilize the kind only access point.

These systems must be able to reserve reports and subscribe to goose or sample values, but they don't need to publish or distribute data using the IEC. Sixty one 850 services, one or more functions that reflect client functionality are present in the client only access point. A client only function often refers to a capacity like a historian or archive my alarm capabilities or some protection functions. These are not logical node classes in interfacing and archiving. Logical Node Group According to the IEC sixty one age 50 standard.

Security

The demand for substation automation is growing in a quickly changing environment. The necessity for cybersecurity and the implementation of IEC, sixty one 850 have been hot concerns in the substation automation systems world. The amount of grid automation and interconnection, as well as I.T. or oti convergence, are ushering in a new age of difficulties for electrical infrastructure. In the meantime, power industry security has become a hot topic on the world stage, particularly that there have been some smart grid breaches happening in the last few years. Interaction with external systems is enabled through the IEC sixty one 850 protocol standard. This enables power grid equipment to interact with one another through standard Ethernet networks.

And as a result of this evolution, energy networks are now vulnerable to the same vulnerabilities that I.T. based systems have. So in this example, we will look at the cyber security aspect of IEC sixty one eight fifty. Now, before we look at the issues, let us take a look at what advantages of IEC sixty one age 50 has first in terms of security. Well, first of all, substation networks in general are separated from corporate networks, and they are usually not connected to a public network or the internet. This physical barrier serves as a first line of defense against a variety of assault situations, and it should be carefully maintained.

Secondly, protocol gateways are frequently used as part of the IEC sixty one age 50 to control the quantity of data that leaves the substation. Because most IEC sixty one is 50 based substations are not directly coupled at the network level. This might be considered an advantage as it provides an additional level of security between the external and internal networks. The protocol gateway serves as a buffer as a result without extensive security testing. This gateway should not be used as a security control. Lastly, the IEC sixty two three five one standard is applied to the sixty one 850 standard to allow Telus encryption to be implemented to the MLS or the manufacturer message specification that I've mentioned earlier in this Book.

Now this has the benefit of allowing the connection to be verified, the channel to be encrypted and or the messages to be signed to assure the communications integrity. However, because the IEC sixty two three five one standard is subject to interpretation on numerous technical aspects and is not yet generally accepted by the industry, the technical implementation of TLS four six one eight 50 may differ across manufacturers and may not always function. So now let us look at the general security weaknesses. IEC six hundred fifty includes many security deficiencies that experienced attackers may exploit to infiltrate the system, perhaps resulting in a blackout of the grid.

Now there are five general ones that I want to touch on before moving on to the specifics, and they are the firmware message, keys, function and authentication. Now let us start with a firmware. There's generally no method to verify the firmware integrity because it isn't usually signed. And this might open the door to more complex hacker exploits scenarios, especially if the supply chain is uncontrolled within the utility company. Now, as for the message protocol, the Guo's protocol has no way of authenticating a publication, so anyone on the network has the ability to impersonate a publisher.

Now, although considerable effort has been made to secure a goose by incorporating a signature, it has demonstrated that the time and performance constraints of goose for protection algorithms make it technically challenging to create a satisfactory solution using the present specification. Thirdly, key management adds another layer of danger since custom made key management infrastructure that doesn't address the correct concerns might leave a system vulnerable while giving the illusion of protection. Now, IEC sixty one 50 includes strong capabilities with its function that may result in unexpected outcomes, but unfortunately these functions are often hard coded within the standard, and as a result, access control levels become extremely difficult to achieve.

Limiting the device's security hardening. Lastly, as mentioned previously, the Advantages chapter although authentication is implemented and MS. The implementation varies a lot depending on the manufacturer's implementation, and in some way they may still be using plain text as a password. Now, some of the MCO files, as you may recall, contain vital system information, and these files must be secured in transition and as well as when it is being stored, according to the IP zero one one Oden and Nurk cybersecurity standards. Now, any files containing operational communication addresses and grid system topology need to be protected, and when these files are at rest, they must be subject to the guide on cybersecurity for data file protection.

And not only will data need to be encrypted, but access control and audit logs will almost certainly be necessary as well. Designing such extra capabilities before implementation might help avert future issues. Now, the IEC sixty two three five one Dash one one specifies procedures to use. The W3C recommendations for XHTML encryption, syntax and processing and XML signature syntax and processing can be implemented to safeguard the contents of files while they are in transit. Now, many capabilities of these W3C specifications are not used by IEC sixty two three five one Dash seven, and as a result, it's preferable to conceive of IEC sixty two three five one Dash 11 as a specifications implementation profile.

Now, the IEC standard is designed to encapsulate, sign and perhaps encrypt exemplar files in their entirety, allowing the wrappers to be removed and regular processing to be done. Now, because in the end, there are simply two zero zero three Dash five and a ship Dash zero zero five Dash five security standards both need access control. IEC sixty one eight fifty Dash 90 Dash 19 describes how role based access control is. RBC can be implemented in a 61 one 850 context. Now, it has been agreed that the security configuration will need to be configured and controlled by someone other than a substation engineer because of the separation of task and function.

This should be a recommended case in general and common sense as well. Now, as a result, the RV AC configuration will be stored in a separate file generated by a security configuration tool. Now, objects in SC Alpha will be referenced by the RBC configuration and to give identities and roles of connected entities. IEC six 2050 will use digital certificate certificates, including attributes certificates because 6.5 o nine certificates are used. The need of a PKI or public key infrastructure is an unintended consequence of this decision, and this involves making use of LDAP or lightweight direct directory access protocol.

Grid operations and dependability are more essential than cybersecurity, which is one of the distinctions between it back and that being worked on inside IEC. Sixty one 50, which is a more cater to words, reliability, reliability and faster response due to the need of critical operations of the IDC. Now, the idea of operating constraints was added into the RBC equations as a result of this to cater towards the applications for the IEC sixty one 850 standard. Now, despite the fact that the IEC sixty one 850 standard enforces a highly standardized and formalized structure, all communications transmitted via the different protocols appear to be susceptible from a cybersecurity standpoint.

There is no method for validating message authenticity, and the data conveyed over these messages is in plain on an encrypted format. Although attack detection and self-healing are not mentioned in sixty one 850 50 documents, a technique such as intrusion detection system or the IDs might be able to be implemented inside the grid to assist security of the IEC sixty one point fifty implementation system. There's a fundamental feature of IEC. Sixty one 850 is interoperable, and apart from its power mapping to TCP IP stack, the standard was mapped over the DWP s web servers to make it easier to communicate with electronic devices inside the grid.

This, however, posed a number of dangers to the grid itself. One of the biggest advantages of IEC sixty one 850 is the usage of goose messages, which ensures high performance operations. Furthermore, owing to Ethernet non deterministic nature, dependability is guaranteed even under high communication load situations. Now, because the original protocol is vulnerable to injection and replay attacks, a malicious person can jeopardize substation security by creating a goose frame that looks like the one seen on a network, but contains altered values to replace the true value of the status being transmitted by the legitimate publisher.

A A personal can insert a cruise control block over 18m substantially larger than existing value and due to the protocols and authenticated nature. There are several versions of this attack. The processing of status numbers and goose frames allows for the implementation of such attacks, and these attacks can be used to hijack the subscriber's communication in order to prevent valid use messages from being processed, as well as to fake further attack traffic in order to influence the subscriber. The impact might be significant depending on the functionality of the specific compromised subscribers.

If the hack subscriber was used to operate electrical protection, for example, its operation might be jeopardized, leading to safety concerns such as great damage or even harm to people. Now there are various types of attack that can be implemented, but here I want to talk about two potential ones, the first one being the high rate flooding attack and the second one being the high status. No attack, no high rate flooding attack. After examining an initial goose frame, the attacker must multicast a range of fixed goose messages with rising status numbers. The high rate flooding of fake packets is intended to use a status number that is greater than the subscribers expected. Status number A status number flooding assault is the best.

We describe this variation. On the other hand, in high status, no attack. After examining goose frames, the attacker multicasts a single fixed goose frame with an extremely high status number to a subscriber. It is intended that after the fake Guo's frame is handled, the subscriber will not process any real goose frames with a status number equal to or less than this. Both the attacks I mentioned are designed to take advantage of the protocols, layers, lack of authentication and confidentiality.

The spoofed goose frames can be handled due to a lack of authentication because there is no confidentiality. The content of Vallecas frames can be read. The attacks are intended to successfully hijack user message communication and hence processing on the subscriber. The subscribers are compelled to process a higher status number than the one used by the legal publisher in these attacks, and as a result, the subscribers are unable to serve genuine goose frames and are thus vulnerable to the attacker's control.

Use of IEC61850 in Power System Protection

Transmission and some transmission systems require high speed unit protection in order to operate efficiently. In the case of a problem, the protective system must immediately isolate the faulty circuit to avoid damage and improve system stability. Two forms of mean protection often used for low protection, and that is current differential and distance protection. Now, the former uses a dedicated signaling channel to convey specific current and vector messages between line ends, while the latter uses tower protection equipment separate from the regular protection relays. But enabling the relay to link directly to the signaling channel accessible between line ends rather than utilizing typical binary input and output connections to signal via other intermediary equipment.

There is significant potential for enhancing the unit protection performance for distance applications. Communication devices, on the other hand, are still frequently utilized. Advanced communication base protection methods can accomplish high speed fall clearance for various problems on tri transmission lines and the overall cost of expedited transmission. Line protection solutions can be significantly reduced thanks to direct relay to relay communications, and at the same time, they minimize the overprotection operating time for any defect inside the protection zone. In this chapter, I am going to look at the use of IEC sixty one 850 in power system protection.

Now, distance protection that is used today does not offer instantaneous shipping for all faults on the transmission line being protected, while communication based accelerated systems allow for a significant reduction in total fault clearing time for any problem inside the zone of protection. They do not have the same high speed communication needs as line differential protection. And this is because a signaling channel is utilized to convey basic on off data in these methods. This gives the remote and protection device more information, which it can utilize to speed up in zone fault clearing or block operation for exterior problems.

Now there are three basic functioning modes for these protection programs. A local protective real operation determines whether or not to deliver a command in each mode, and the three modes are into tripping, permissive and blocking. Let's start off with intercropping. Instruction is not monitored by any protective mechanism at the receiving end in intercropping mode and merely triggers a Precor trip action because the received signal was not checked. It is critical that any noise on the signal and channel is not misinterpreted as a genuine signal, or, to put it another way and into the shipping channel, must be extremely safe. Next is the permissive mode.

Tripping is only allowed in permissive function when a command coincides with a protective action on the receiving end. The signaling channel for permissive schemes does not need to be as secure as into tripping channels because a second independent check is performed before tripping. And lastly, we have the blocking function. Remote tripping is only allowed in blocking mode when no signal is received, but a protective procedure has happened. In other words, even a protective operation happens when a command is delivered. The receiving end device is prevented from acting because the signal is utilized to prevent tripping.

It is obvious that it is received whenever and as fast as possible. To put it another way, a blocking channel must be quick and reliable. While the kind of scheme employed is determined by the protection function that delivers the permissive or blocking signal to the remote end, and when it comes to distance, we generally refer to permissive underreaching, overreaching or blocking schemes. We have directional comparison schemes where a directional element is utilized to begin the transmission of a signal to the protected lines. Remote and permissive or blocking directional comparison systems exist with directional components, starting signal transmission and providing monitoring at the receiving end.

Goose messages are used for peer to peer communications in the IEC sixty one 850 integrated substation protection and control system during the setup phases of the substation integration process. They employ multicast ethernet communications to represent the asynchronous recording of an IED functional element, change of state to other peer devices and roll to receive it as communications are utilized to replace the hardwired control signals exchanged between IEDs for interlocking and protection reasons. And as a result, their mission is critical, time sensitive and must be extremely dependable.

The goose message is not a command in a sense that it does not direct any receiving device. What to do. It simply states that a new event has occurred together with the nature of the event and the time it occurred. The information provided in the message is used by the IoT to calculate the appropriate reaction for the given change of condition. Local intelligence in the IED receiving mixed messages determines the needed action to goose messages and what to do if a message runs out due to a communication breakdown. The standard defines high speed peer to peer communications as interface if eight or direct data flow across bays, particularly for rapid operations like interlocking or protection.

IEC Sixty one 850 establishes highly stringent performance criteria due to the importance of the activities performed viscous messages. The concept is that high speed peer to peer connections should be as good as or better than what is now possible with present technologies, and as a result, the overall peer to peer times should not be more than four seconds. And this is the entire time between the publishing devices functional element and the subscribing devices information receival. Now, another essential need for good communications that they need to be extremely reliable, given the importance of a message like break or failure protection function or fault and reverse direction, and the fact that the messages are not verified by multicast, that there must be a system in place to guarantee that the receiving ideas get the message and operates as intended.

Messages are repeated as long as the state continues to ensure a high level of dependability, a communication with a period to live referred to as a whole time in order to maximize reliability and security. So unless the same status messages repeat or new messages received before the whole time and the message status will expire when the whole time expires, the initial goose message has a small repetition duration and successive messages have longer repeat and whole times until a maximum is achieved. The goose message includes information that lets the receiving IED know if a message was missed in a status has changed and how long it has been since the last status change.

The goosed message is based on the notion of having a single message that provides all needed protection scheme information for an individual protection IED in order to achieve high speed performance, while also reducing network traffic under severe fault circumstances. It depicts a state machine that informs its peers about IED status and depending on the substation architecture and the kind of protected unit, high speed peer to peer communications can be used in various distributed protection methods. Goose messages and push some of the most sophisticated capabilities of newer versions of Ethernet to increase the security and performance of protection systems across the substation local area network on Ethernet networks.

The Virtual Local Area Network, or the VLAN protocol, allows an identifier or tag to be inserted into the Ethernet frame format to identify the VLAN to which the frame belongs to. It allows frames from devices to be assigned to logical groups, which has a number of advantages, including improved network security. Another use of a higher priority for goose messages allows distributed protection schemes overall performance to be improved, especially during periods of high network traffic, such as when a problem occurs. Villain IDs can help to improve the security of this distributed scheme implementation.

Conclusion

Congratulations for you having just completed this Book right now. You should have learned the IEC sixty one 850 data model. The concept of the generic substation events such as the Ghost Protocol, as well as the manufacturing message specification. Last but not least, you have also looked at the security features and issues pertaining to the IEC sixty one 850 standard. So what will be your next step? Well, you should start applying the concepts that you learn from this Book in your career. If you are a skilled engineer and if you are or your company is implementing IEC sixty one eight fifty in your scalar systems. Well, congratulations, because now you have the best opportunity to apply what you learn in this Book in your career.

Now I would also recommend you to actually get a copy of the IEC sixty one 850 standard, as I've mentioned before. It is a very meaty standard with a lot of documentation, so I have not actually covered every single bit and piece of this standard. Now get the standard and read it. I think it would actually get you from being a beginner in looking at just the fundamentals to being a professional in the IEC sixty one 850 standard. So be sure to actually get a copy of the standard and read it through to make sure that you at least understand.

And the concept, the basic concept around this standard, as well as just a brief idea as to where you can find this stuff in your standard so that when you need to look stuff up, you can actually reference the standard fast and efficiently. Lastly, if you're interested in the different electrical engineering knowledge pertaining to the utility, energy and renewable energy industry, I recommend you to visit my other Books as I have also provided various types of fundamental knowledge in various parts of the concepts within electrical engineering, within the energy and the utility industry. Well, I wish you good luck in your career and I hope to see you in my other Books. Thank you.

www.ingramcontent.com/pod-product-compliance
Lightning Source LLC
Chambersburg PA
CBHW071952210526
45479CB00003B/911